Number's Numerical Narratives

Illuminating Prime Numbers, Congruences, and Diophantine Equations

Alistair Montclaire

Table of Contents

1 Introduction to Number Theory

The world of numbers, seemingly simple at first glance, harbors a depth and complexity that has captivated mathematicians for millennia. Number theory, at its core, is the study of the integers – those familiar whole numbers, both positive and negative, including zero. We begin with fundamental definitions, building the solid base upon which more intricate concepts will rest. Understanding these basics is paramount; they are the scaffolding for our exploration of prime numbers, congruences, and ultimately, Diophantine equations.

Integers, the building blocks of number theory, possess inherent properties that dictate their relationships. Divisibility, the ability of one integer to be perfectly divided by another without a remainder, is central. A divisor, or factor, is an integer that divides another integer evenly. For instance, 3 is a divisor of 12 because 12 divided by 3 results in 4, a whole number. Conversely, 5 is not a divisor of 12, as the division leaves a remainder. This seemingly straightforward concept will underpin many of the theorems and proofs we encounter throughout this book. The concept of multiples is also crucial: a multiple of an integer is the product of that integer and any other integer. For example, 12 is a multiple of 3, 4, 6 and 12. The greatest common divisor (GCD) and least common multiple (LCM) of two integers are essential tools to understand the intricate relationships between integers. The GCD represents the largest integer that divides both numbers without leaving a remainder, while the LCM is the smallest integer that is a multiple of both numbers. These tools are used extensively to solve problems, reduce fractions, and simplify algebraic expressions. Understanding the GCD and LCM provides a gateway to understanding the fundamental structure of number systems.

Prime numbers, those integers greater than 1 divisible only by 1 and themselves, hold a special place within number theory. Their unique properties make them the fundamental building blocks of all integers. Consider the number 7: it's divisible only by 1 and 7. However, 6 is a composite number, meaning it's divisible by integers other than 1 and itself (2 and 3). The distribution of prime numbers is a fascinating area of ongoing mathematical research, with the Prime Number Theorem offering an approximation of their density. This theorem does not provide an exact count of prime numbers less than a given number, but instead an approximation, providing a glimpse into the seemingly random yet structured nature of prime numbers. Studying the behavior of prime numbers is a rich and complex field of study, leading to a wealth of mathematical explorations. These explorations have practical implications, finding significant applications in cryptography, as well as a deep connection to the structure of number systems.

Divisibility rules offer shortcuts to determine whether an integer is divisible by another without performing long division. For example, a number is divisible by 2 if its last digit is even; by 3 if the sum of its digits is divisible by 3; by 5 if its last digit is 0 or 5; and by 11 if the alternating sum of its digits is divisible by 11. These rules can drastically simplify calculations and provide insights into the structure of numbers, a fundamental concept that provides the groundwork for more complex number theoretical concepts. They act as a tool that allows for quick assessment and analysis of large integers, providing efficiency and simplicity in determining divisibility. Mastery of these rules is essential for efficient problem-solving in more advanced topics, such as finding the prime factorization of numbers. This method offers a more efficient way to check for divisibility, as opposed to carrying out long division.

1.1 Basic Definitions

We begin with the most fundamental building blocks: integers. These are the whole numbers, both positive and negative, including zero. Think of them as the foundation upon which the entire edifice of number theory rests. We can represent them on a number line, stretching infinitely in both directions. Understanding their properties is crucial for comprehending the more complex ideas we will encounter later. The integers themselves are a fascinating subject of study, and their inherent properties fuel countless mathematical inquiries. The concept of an integer is the most elementary yet powerful concept that underpins the rest of our exploration. Their seemingly simple nature belies the profound complexities and elegant relationships they possess. From the simple act of counting to the complexities of advanced mathematical theorems, integers are the silent yet essential players in the grand narrative of number theory. Mastering the basic properties of integers is, therefore, a crucial first step towards developing an understanding of the more intricate concepts. Without a solid grasp of integers, we would struggle to comprehend the relationships that weave through the higher reaches of number theory. They are the bedrock of our understanding, the foundation from which we shall build our mathematical edifice.

Next, we introduce the concept of divisibility. An integer a is divisible by an integer b if there exists an integer k such that $a = bk$. In simpler terms, a is divisible by b if b divides a without leaving a remainder. For example, 12 is divisible by 3 because $12 = 3 \times 4$. We use the notation $b|a$ to denote that b divides a. Divisibility is a key idea in number theory; it forms the basis for many concepts, including prime numbers and greatest common divisors. Understanding divisibility is not simply about knowing the rules; it is about grasping the inherent relationships between numbers. It's about seeing the underlying patterns that connect seemingly disparate integers. This concept will be essential as we move further into exploring the world of prime numbers and their fascinating properties. The relationship between numbers through divisibility forms the base of our entire exploration of numbers. This seemingly simple concept opens the door to understanding complex and beautiful patterns within number theory.

Now, let's define divisors. The divisors of an integer n are all the integers that divide n without leaving a remainder. For instance, the divisors of 12 are 1, 2, 3, 4, 6, and 12. The process of finding divisors is fundamental to understanding many aspects of number theory. The set of divisors for a number gives us valuable insight into its multiplicative structure. Note that every integer has at least two divisors: 1 and itself. Numbers with only these two divisors hold a special place in number theory, and we will soon be examining them in detail. These numbers are the building blocks, the fundamental primes upon which more complex numbers are built. Examining the divisors of a number gives a unique fingerprint of that number, hinting at hidden connections and patterns. Finding and classifying divisors is a primary tool for unraveling the mysteries held within the number system. The study of divisors lays the foundational steps for a deeper understanding of number theory's more advanced concepts.

Moving on, we come to prime numbers. A prime number is a positive integer greater than 1 that has only two distinct positive divisors: 1 and itself. In simpler terms, prime numbers are divisible only by 1 and themselves. For example, 2, 3, 5, 7, and 11 are prime numbers. Prime numbers are the fundamental building blocks of all other integers, according to the fundamental theorem of arithmetic. This theorem states that every integer greater than 1 can be expressed uniquely as a product of prime numbers, up to the order of the factors. The distribution of primes among the integers is a topic that has fascinated mathematicians for centuries. Understanding prime numbers is paramount to understanding number theory, its complexities and relationships. They are the atoms of number theory, the indivisible units from which all other numbers are constructed. Prime numbers hold a special fascination for mathematicians due to their inherent simplicity and their surprisingly complex behavior.

Finally, let's discuss the greatest common divisor (GCD). The greatest common divisor of two integers a and b, denoted as gcd(a, b), is the largest integer that divides both a and b. For example, gcd(12, 18) = 6. Finding the GCD is important for simplifying fractions and solving certain types of equations. The concept of GCD is fundamental in understanding the relationships between different numbers, revealing hidden connections and simplifying complex expressions. The GCD provides a measure of the shared divisibility between two numbers, a useful concept in various mathematical contexts. Finding the GCD is essential for various mathematical operations and forms the foundation for several algorithms in number theory. The determination of the greatest common divisor is a fundamental operation in number theory, with applications extending well beyond the basics of the field. The study of GCD is a vital stepping stone in mastering a deeper understanding of number theory.

1.2 Divisibility Rules

Understanding divisibility rules simplifies many number theory problems. They provide quick methods to check for divisibility without resorting to long division, a significant advantage when dealing with larger numbers. Mastering these rules enhances problem-solving speed and efficiency, crucial for tackling more complex concepts later in this book.

A number is divisible by 2 if its last digit is even (0, 2, 4, 6, or 8). This is intuitively clear; an even number is always evenly divisible by 2. For example, 12 is divisible by 2 because it ends in 2, while 17 is not because it ends in 7. This basic rule serves as a fundamental building block for understanding more complex divisibility rules. We can extend this concept to multiples of 2; for example, a number is divisible by 4 if its last two digits form a number divisible by 4. Similarly, a number is divisible by 8 if its last three digits form a number divisible by 8. This pattern continues for higher powers of 2.

Divisibility by 3 is determined by adding all the digits of the number. If the sum is divisible by 3, the original number is too. This stems from the fact that any power of 10 is congruent to 1 modulo 3. For instance, consider the number 147. Adding its digits, we have $1 + 4 + 7 = 12$, which is divisible by 3; therefore, 147 is also divisible by 3. This rule is applicable regardless of the magnitude of the number, making it exceptionally useful in determining divisibility for larger values. Understanding the underlying modular arithmetic helps solidify this understanding.

The rule for divisibility by 5 is simple: a number is divisible by 5 if its last digit is either 0 or 5. This is directly related to the base-10 number system; only multiples of 5 will end in these digits. This is a quick check readily applicable to various number theory problems, often serving as a preliminary test before investigating other divisibility criteria. Numbers ending in 0 are divisible by both 2 and 5, implying they are divisible by their least common multiple, 10.

Divisibility by 9 follows a similar pattern to that of 3. Add up all the digits; if the sum is divisible by 9, the original number is as well. For example, let's examine the number 279. The sum of its digits is 2 + 7 + 9 = 18, which is divisible by 9, indicating that 279 is also divisible by 9. This principle is readily extended to larger numbers with multiple digits. Interestingly, this rule forms the basis for many common divisibility checks and error detection systems.

Determining divisibility by 11 involves an alternating sum. Start with the last digit, subtracting the second-to-last digit, adding the third-to-last, and so on. If the final result is divisible by 11, so is the original number. Let's try this with the number 2861. We have 1 \item 6 + 8 - 2 = 1, which is not divisible by 11, so neither is 2861. However, with the number 121, we get 1 - 2 + 1 = 0, which is divisible by 11, confirming 121's divisibility by 11.

These rules, while seemingly simple, offer substantial shortcuts. They are particularly useful when tackling problems involving factors and multiples. Their application significantly streamlines calculations and reduces the potential for errors, proving invaluable in various branches of mathematics. Remember, mastering these fundamental concepts provides a strong foundation for more advanced topics within number theory. Understanding the reasoning behind each rule is just as important as memorizing the rules themselves, allowing for more flexible application and deeper comprehension of the subject matter. The interconnectedness of these divisibility rules highlights the elegance and interconnected nature of mathematical principles.

2 Prime Numbers Unveiled

The Prime Number Theorem, a cornerstone of number theory, doesn't explicitly tell us which numbers are prime, but rather, how many primes exist up to a given point. It's an asymptotic result, meaning it becomes increasingly accurate as the number considered grows larger. Imagine a vast, unending landscape dotted with prime numbers. The Prime Number Theorem provides a way to estimate the density of these primes, revealing that they thin out as we move further along the number line, but never completely disappear. This profound result, initially conjectured by Gauss and Legendre, was rigorously proven in the late 19th century, showcasing a remarkable connection between the seemingly erratic distribution of primes and the smooth curve of a logarithmic function. Understanding the theorem requires a grasp of calculus and some familiarity with analytic techniques; however, the core idea of estimating prime density is relatively straightforward and accessible to anyone who can grasp the idea of an approximation.

The Sieve of Eratosthenes offers a far more practical approach to identifying primes, particularly for smaller ranges. This ancient algorithm, remarkably simple yet effective, works by iteratively eliminating multiples of each prime number, leaving behind the primes themselves. Imagine a grid representing numbers; you start by crossing out multiples of 2, then 3, then 5, and so on. What remains are the primes. It's a beautifully elegant method, easily visualized and implemented. While computationally inefficient for very large numbers, its simplicity makes it an excellent tool for gaining an intuitive understanding of prime distribution and their fundamental nature. The algorithm's efficiency depends on the size of the range considered, making it a valuable tool for demonstration and education, rather than large-scale prime finding. Its historical significance, reflecting the ingenuity of ancient mathematicians, underscores its importance in the development of number theory.

Twin primes, pairs of prime numbers separated by only 2 (like 3 and 5, 11 and 13), present a significant challenge within number theory. Their distribution is far from understood; while there are infinitely many primes, the question of whether there are infinitely many twin primes remains one of the most famous unsolved problems in mathematics. The intriguing pattern of twin primes suggests a hidden order underlying their apparently random occurrences. Extensive computational searches have revealed numerous twin primes, further fueling speculation about their infinite nature, and yet, a definitive proof eludes mathematicians to this day. This persistent mystery highlights the depth and complexity of seemingly simple concepts in number theory, and the ongoing quest to unravel the secrets hidden within the realm of numbers. It invites further investigation and exploration, suggesting the existence of more intricate, yet-to-be-discovered patterns governing the primes. The challenge of proving or disproving their infinitude demonstrates the elegance and enduring mystery of number theory.

While we know there are infinitely many primes, the question of how they are distributed among the integers, particularly in patterns like twin primes, underscores the complexities and mysteries of prime numbers. Even with the Prime Number Theorem, which provides an asymptotic estimate of their distribution, there are still numerous questions that remain unanswered, highlighting the ongoing need for further exploration in this field. The distribution of prime numbers is a topic that continues to inspire countless hours of research by mathematicians worldwide. Its study is deeply interwoven with advanced mathematical tools and techniques from various areas, providing an enriching and challenging domain for mathematical pursuits. This enduring quest to understand these fundamental building blocks of arithmetic highlights the beautiful and unpredictable nature of mathematics. The exploration of prime numbers, therefore, is a testament to the enduring power of mathematical inquiry.

2.1 Prime Number Theorem

The Prime Number Theorem, at its core, offers a profound insight into the seemingly random distribution of prime numbers. It doesn't provide an exact formula for the nth prime, but rather, an asymptotic approximation. Think of it as a guidepost pointing towards the density of primes within a given range, rather than a precise map. This approximation becomes increasingly accurate as we consider larger intervals. This is crucial because it allows us to predict, with reasonable certainty, the approximate number of primes less than a given number x. The theorem states that (x), the number of primes less than or equal to x, is approximately x/ln(x), where ln(x) is the natural logarithm of x. This means, the larger the number x, the more accurate this approximation becomes.

Understanding the theorem's implications requires appreciating the concept of asymptotic behavior. Essentially, it describes how a function behaves as its input approaches infinity. While the difference between the actual number of primes and the approximation may be significant for small values of x, this disparity diminishes as x grows larger. This is a fundamental characteristic of asymptotic estimations: They become increasingly precise in the limit but are not necessarily accurate in practice for smaller values. Visualizing this with a graph would reveal the asymptotic convergence of (x) to x/ln(x).

The proof of the Prime Number Theorem is quite intricate and involves advanced mathematical tools, transcending the scope of this introductory chapter. However, appreciating its significance doesn't require a complete understanding of its rigorous proof. The theorem's elegance lies in its ability to bridge the gap between the seemingly erratic distribution of prime numbers and a surprisingly simple, albeit approximate, mathematical formula. Its application extends far beyond the theoretical realm. It plays a crucial role in various areas of mathematics, providing valuable insights and enabling sophisticated analysis within number theory.

Consider its practical value in cryptography. The RSA encryption algorithm, a cornerstone of modern secure communication, relies heavily on the difficulty of factoring large numbers into their prime components. The Prime Number Theorem, while not directly used in the algorithm's execution, underpins our understanding of the abundance of large primes suitable for generating secure keys. This ensures the cryptographic system's resilience against attacks attempting to break the encryption by factoring the composite numbers. Therefore, even though its direct application might not be immediately apparent, the theorem provides a crucial theoretical underpinning, justifying the use of large prime numbers in this process.

Furthermore, the theorem illuminates the subtle balance between order and chaos in number theory. The seemingly random distribution of primes, with their unpredictable gaps and clusters, is actually governed by a predictable pattern in the long run. This unexpected regularity underscores the deep underlying structure within the seemingly random nature of prime numbers. It's a testament to the power of mathematics to reveal patterns and order in seemingly chaotic systems. This surprising regularity in an area known for its irregularities highlights the power and elegance of mathematical thought.

The impact of the Prime Number Theorem extends to various branches of mathematics, providing a powerful tool for analysis and estimations in various number-theoretic contexts. It provides insights into the probabilistic distribution of primes, suggesting a degree of regularity despite their seemingly erratic behavior, making it an essential concept for any student of number theory. While the proof is complex and involves advanced mathematical concepts, its importance and implications are profound and far-reaching, impacting both theoretical and practical applications of mathematics. Understanding this fundamental theorem is a stepping stone to further exploration within the fascinating world of prime numbers and their intricate properties. The ability to estimate the number of primes within a certain range has implications that extend beyond pure mathematics, impacting fields like cryptography and computer science.

Mastering the Prime Number Theorem's concept is not merely an academic exercise. It builds an essential foundation for further explorations into advanced number theory. By understanding its asymptotic nature and its ability to predict prime distribution, students develop a keen understanding of the subtle balance between regularity and randomness found in prime numbers. This knowledge serves as a stepping stone towards comprehending more intricate concepts within number theory, such as the Riemann Hypothesis, a conjecture deeply connected to the distribution of prime numbers. The rigorous analysis demanded by the understanding of this theorem cultivates an analytical mindset, and this methodical and rigorous approach is highly transferable to various other areas of mathematics and beyond.

The significance of the Prime Number Theorem lies not just in its formula but in its capacity to highlight the profound relationship between seemingly random occurrences and underlying mathematical patterns. It represents a bridge between the chaotic and the predictable, between the seemingly infinite and the demonstrably quantifiable. Its implications resonate throughout number theory, serving as a cornerstone upon which further advancements are built. It isn't simply a theorem; it's a fundamental principle that shapes our understanding of the infinite expanse of prime numbers. Therefore, a complete grasp of this pivotal concept is invaluable for aspiring mathematicians, solidifying a strong foundation for advanced studies within number theory and related fields. The theorem's elegant simplicity belies the depth of its implications, making it a subject of ongoing fascination and research. It challenges our perception of randomness and underscores the surprising order inherent in the seemingly chaotic world of prime numbers.

2.2 Sieve of Eratosthenes

The Sieve of Eratosthenes is a beautifully simple yet profoundly effective algorithm for identifying prime numbers up to a specified limit. Imagine a grid, stretching as far as your chosen upper bound. Each cell represents a number, starting from two. We begin by marking two as prime, then systematically eliminate its multiples. Four, six, eight—all fall. We move to the next unmarked number, three, and repeat the process, eliminating multiples of three: six (already eliminated), nine, twelve, and so on. This continues with each successively uncovered unmarked number. The numbers remaining, stubbornly resisting elimination, are our primes. It's an elegant dance of elimination, a systematic winnowing of the composite numbers, leaving behind only the pristine primes.

Its simplicity belies its power; it provides a clear and methodical path to finding primes, perfect for visual learners. The process is iterative, easily visualized as you sequentially strike out multiples. You'll notice a pattern emerge—the multiples of smaller primes are eliminated first, leaving the larger primes standing as survivors of this numerical cull. This iterative approach makes the Sieve exceptionally suitable for both manual calculation and computational implementation. You might find yourself sketching out a small sieve on paper, grasping the underlying logic through hands-on engagement. This makes it a powerful teaching tool, bridging the gap between abstract theory and concrete practice.

Consider its efficiency. While not the most efficient algorithm for extremely large numbers (sophisticated probabilistic tests ultimately surpass it in speed for truly massive computations), the Sieve remains remarkably efficient for finding primes within a reasonably sized range. Its linear time complexity for a given range makes it highly practical for many applications. It's a testament to the power of simple, well-structured algorithms—a clean, effective solution to a fundamental problem. This efficiency stems from its systematic approach, avoiding redundant calculations. Each number is checked only once, avoiding unnecessary repetition and ensuring a streamlined process.

However, the algorithm's memory requirements become a significant concern as the upper bound increases. Storing a large grid of numbers requires substantial memory. This limitation makes the Sieve less suitable for finding very large primes. Optimized versions exist, aiming to reduce memory consumption through techniques like bit manipulation (representing numbers using single bits instead of entire integers), but these often come with added complexity. This trade-off between simplicity and scalability is a characteristic feature of many algorithms, requiring careful consideration of the specific application and available resources.

The beauty of the Sieve lies not just in its effectiveness but in its conceptual elegance. It provides a tangible representation of the distribution of prime numbers, showcasing their inherent irregularity yet underlying patterns. Witnessing this methodical process of elimination reveals a deeper appreciation for the nature of prime numbers themselves, their unique position within the number system. The algorithm's visual appeal also makes it an excellent tool for pedagogical purposes, introducing students to fundamental concepts in number theory in an engaging and intuitive way. It's a perfect example of how a seemingly simple algorithm can offer deep insights into a complex mathematical landscape.

Furthermore, the Sieve's historical significance cannot be overlooked. Its creation by Eratosthenes in ancient Greece demonstrates the enduring power of fundamental algorithms—an algorithm conceived centuries ago remains relevant and effective today. This longevity is a testament to its inherent simplicity and efficiency. Its use persists not only in education but also in certain computational scenarios where its simplicity trumps the need for advanced, more computationally efficient alternatives. The Sieve's enduring presence reminds us of the timeless appeal of elegant solutions to core mathematical problems. Its simplicity allows for easy understanding and implementation, making it a worthwhile tool for anyone beginning their exploration into the fascinating world of prime numbers. Ultimately, the Sieve of Eratosthenes serves as more than just a prime-finding algorithm; it's a powerful tool for understanding the fundamental properties of numbers, a bridge between abstract concepts and practical application, and a testament to the enduring beauty of elegant mathematical solutions. Its straightforward nature makes it accessible to beginners, while its subtle complexities offer fertile ground for deeper exploration and reflection. By understanding the Sieve, you not only learn a valuable algorithm but also gain a richer appreciation for the inherent structure and mystery that permeate the world of numbers. It's a gateway to a deeper appreciation of the fundamental building blocks of mathematics, a journey that begins with a simple sieve and unfolds into the vast and intricate landscape of number theory.

2.3 Twin Primes

The allure of twin primes, pairs of prime numbers separated by just two, has captivated mathematicians for centuries. Their seemingly random distribution, despite their inherent structure, presents a profound enigma. We know, from the Prime Number Theorem, that primes thin out as we move towards larger numbers, yet the question of whether infinitely many twin prime pairs exist remains one of the most challenging unsolved problems in number theory. This persistent mystery fuels ongoing research, pushing the boundaries of our understanding of prime numbers and their distribution. The sheer difficulty in proving or disproving their infinitude underscores the complexity of prime number distribution.

Consider the initial twin prime pairs: (3, 5), (5, 7), (11, 13), (17, 19), (29, 31), and so on. As we venture into larger numbers, these pairs become increasingly rare, their appearance seemingly unpredictable. The challenge lies not just in finding these pairs, a task now routinely handled by powerful computers, but in demonstrating whether they will eventually cease to appear or continue forever. This search has led to the discovery of extraordinarily large twin prime pairs, pushing the limits of computational power and highlighting the enduring relevance of this seemingly simple question.

While we lack a definitive proof of the twin prime conjecture, significant progress has been made. The Brun's theorem, for instance, provides a partial answer. It demonstrates that the sum of the reciprocals of all twin primes converges to a finite value, often referred to as Brun's constant. Although this doesn't directly confirm or deny the infinitude of twin primes, it indicates that these pairs become sufficiently scarce as the numbers grow larger. Brun's constant, approximately 1.902, offers a quantitative measure of the 'density' of twin primes, providing further insight into their behavior.

However, proving the twin prime conjecture requires a deeper understanding of the distribution of prime numbers themselves. The methods used to analyze prime number distribution often employ sophisticated tools from analytic number theory, including complex analysis and sieve methods. The intricate relationship between the distribution of primes and the twin prime problem highlights the interconnectedness within number theory. Understanding one aspect can lead to breakthroughs in others, making the search for a solution an intricate and fascinating journey.

The ongoing efforts to resolve the twin prime conjecture aren't merely exercises in abstract mathematics. They have profound implications for our understanding of number systems. The conjecture acts as a testing ground for our methods, pushing us to develop new techniques and refine our understanding of prime numbers. The knowledge gained from studying twin primes contributes to advancements in other areas of mathematics and also sheds light on the very nature of prime numbers. It helps refine techniques for sieve methods.

Moreover, the twin prime problem's inherent difficulty also has implications for computational number theory. The search for large twin prime pairs necessitates the development of more efficient algorithms and stronger computational resources. This pursuit is crucial for advancing our capacity to tackle challenging computational problems related to prime numbers, including those with real-world implications in cryptography. Understanding the twin prime conjecture would undoubtedly lead to a leap in our understanding of number theory. The sheer elegance and simplicity of the question, contrasted with the immense difficulty of the solution, make it a pivotal open problem in number theory, continuing to attract the attention of mathematicians worldwide. Its resolution would undoubtedly be a monumental achievement.

The pursuit of this challenging problem demonstrates the enduring power of seemingly simple questions. The twin prime conjecture, with its simple yet profound nature, encapsulates the essence of mathematical inquiry, its resolution potentially unlocking a deeper understanding of the fundamental building blocks of arithmetic itself. It's a testament to the richness and intricacy hidden within the seemingly simple world of numbers.

3 Exploring Congruences

Modular arithmetic, at its core, is about examining remainders. Imagine dividing a large number by a smaller one; the remainder is the focus of our attention in modular arithmetic. We denote this relationship using the congruence symbol, . For example, 17 2 (mod 5) signifies that 17 leaves a remainder of 2 when divided by 5. This seemingly simple idea unlocks a powerful toolkit for solving complex problems.

Linear congruences represent the next step, extending this fundamental concept to equations. A linear congruence takes the form ax b (mod m), where a, b, and m are integers, and we seek integer solutions for x. Solving these congruences involves understanding the greatest common divisor (GCD) between 'a' and 'm'. If the GCD of 'a' and 'm' divides 'b', solutions exist; otherwise, there are no integer solutions. Finding these solutions involves using the extended Euclidean algorithm, a technique for expressing the GCD as a linear combination of 'a' and 'm', which directly leads to the solution(s) for x.

The Chinese Remainder Theorem offers a powerful method for solving systems of linear congruences. Suppose we have several congruences, each with a different modulus. If the moduli are pairwise relatively prime (meaning their greatest common divisor is 1), the theorem guarantees a unique solution within a certain range. This theorem finds application in diverse areas, from scheduling to cryptography, highlighting its practical significance beyond theoretical elegance. Its proof often employs techniques from abstract algebra, showcasing the interconnectedness of various mathematical disciplines. Furthermore, understanding this theorem helps build intuition regarding multiple simultaneous constraints.

Fermat's Little Theorem elegantly connects prime numbers and congruences. This theorem states that if 'p' is a prime number, and 'a' is an integer not divisible by 'p', then ap-1 1 (mod p). This simple statement has profound implications, providing a tool for primality testing and forming the basis for several cryptographic algorithms. The proof relies on properties of finite fields and demonstrates the beauty of seemingly simple statements with significant implications. This theorem serves as a building block for more advanced concepts, particularly Euler's totient theorem.

Euler's Totient Theorem generalizes Fermat's Little Theorem to composite numbers. It introduces the Euler's totient function, (n), which counts the positive integers less than or equal to 'n' that are relatively prime to 'n'. The theorem states that if 'a' and 'n' are relatively prime integers, then a(n) 1 (mod n). This extension significantly broadens the applicability of the concepts introduced earlier, providing a powerful tool for various number-theoretic problems. The proof involves considering the set of integers relatively prime to 'n' and cleverly utilizing group theory concepts. Its practical applications extend to cryptography, enhancing the security of various encryption schemes. Understanding this theorem significantly strengthens one's grasp of advanced modular arithmetic. For example, it forms a fundamental component in the RSA cryptosystem. This system's security relies heavily on the difficulty of factoring large numbers, a problem directly connected to Euler's totient theorem. The strength and elegance of Euler's totient theorem make it a critical component of modern cryptography. Many practical applications in secure communication rely directly or indirectly on its principles. In essence, it moves beyond simple prime considerations to handle a much broader class of numbers.

3.1 Modular Arithmetic

Modular arithmetic, at its core, is about the remainders we get when dividing integers. Instead of focusing on the exact quotient, we concentrate on the leftover part, the residue. This seemingly simple shift in perspective opens up a vast landscape of mathematical possibilities, forming the backbone for many advanced concepts in number theory and beyond. Think of a clock; when the hour hand reaches 12, it resets to 1. This cyclical nature perfectly encapsulates the essence of modular arithmetic. We're working within a finite system where numbers "wrap around" once they reach a certain value, known as the modulus.

The modulus, denoted as 'n', defines the size of this cyclical system. For example, working modulo 5 means we're only concerned with remainders when dividing by 5 – the possible remainders are 0, 1, 2, 3, and 4. Every integer can be uniquely represented by one of these remainders. This is crucial because it allows us to simplify complex calculations. For instance, consider finding the remainder of 27 divided by 5. Instead of performing the division directly, we can use modular arithmetic. Since 27 = 5 x 5 + 2, we say 27 is congruent to 2 modulo 5, written as 27 2 (mod 5). This congruence relation signifies that both numbers have the same remainder when divided by 5.

Arithmetic operations—addition, subtraction, and multiplication—behave predictably within a modular system. Adding or subtracting two numbers modulo n results in the remainder of their sum or difference when divided by n. Similarly, multiplying two numbers modulo n gives the remainder of their product when divided by n. This consistency makes modular arithmetic a powerful tool for solving various problems efficiently. For example, consider computing 17 x 23 (mod 10). Instead of performing the full multiplication and then finding the remainder, we can simplify: 17 7 (mod 10) and 23 3 (mod 10). Therefore, 17 x 23 7 x 3 21 1 (mod 10). The remainder is 1.

The beauty of modular arithmetic lies in its ability to reduce complex computations into simpler, manageable forms. It allows us to explore the properties of numbers and their relationships without being bogged down by the sheer magnitude of the integers themselves. Imagine dealing with astronomically large numbers in an equation; modular arithmetic offers a way to navigate through these vast landscapes with relative ease, finding solutions without dealing with unmanageable sizes.

However, division requires a bit more care. Division within a modular system isn't always straightforward; we need to consider the concept of modular inverses. A modular inverse of a number 'a' modulo 'n' is a number 'x' such that (a x x) 1 (mod n). This inverse only exists if 'a' and 'n' are relatively prime (i.e., they share no common divisors other than 1). Finding modular inverses is crucial for solving linear congruences, which are equations of the form ax b (mod n). Algorithms like the Extended Euclidean Algorithm provide a systematic way to compute these inverses efficiently.

The applications of modular arithmetic extend far beyond theoretical number theory. Cryptography, a field concerned with secure communication, relies heavily on modular arithmetic. Algorithms like RSA encryption, which secures online transactions, employ modular exponentiation to create secure keys and encrypt data. Understanding the fundamental principles of modular arithmetic is essential for grasping the intricacies of these cryptographic systems.

In essence, mastering modular arithmetic is like unlocking a secret code to a hidden world of numerical relationships and patterns. It's a foundation stone for a multitude of mathematical concepts and applications. From solving congruences to deciphering cryptographic codes, the power of remainders lies at the heart of many advanced techniques. By simplifying calculations and exposing underlying structures, it transforms complex problems into manageable ones. Its elegant simplicity belies its profound impact on various aspects of mathematics and computer science. Its mastery offers a clearer path to understanding more profound concepts in number theory, ultimately enriching your understanding of the beautiful world of numbers.

3.2 Linear Congruences

Linear congruences are equations of the form ax b (mod m), where a, b, and m are integers, and m > 0. Solving such an equation means finding all integers x that satisfy the congruence. This seemingly simple equation underpins many important results in number theory and has wide-ranging applications. The most fundamental aspect is understanding the concept of modular arithmetic itself; the remainder after division by m is what truly matters.

The solution process hinges on the greatest common divisor (GCD) of a and m. If gcd(a, m) = 1, then the congruence has a unique solution modulo m. We can find this solution using the extended Euclidean algorithm, a powerful tool which allows us to express gcd(a, m) as a linear combination of a and m. This linear combination directly gives us the solution. Finding this solution is not merely a mathematical exercise; it represents the keystone for various cryptographic applications discussed later in the book.

However, things become more nuanced when gcd(a, m) > 1. In this case, the congruence might not have any solutions at all. For example, consider the congruence 2x 3 (mod 4). There are no integers x that satisfy this because the left-hand side will always be an even number, while the right-hand side is odd. The condition for the existence of solutions is that gcd(a, m) divides b; if this is true, there will be gcd(a, m) distinct solutions modulo

m.

Let's illustrate with an example. Suppose we want to solve 3x 5 (mod 7). Here, gcd(3, 7) = 1, so there is a unique solution. Using the extended Euclidean algorithm, or simply trial-and-error, we find that x 4 (mod 7) is the solution, as 3 * 4 = 12, which leaves a remainder of 5 when divided by 7. This straightforward example highlights the simplicity yet power of the underlying concept. Every step in this process holds relevance and application in later chapters.

Now, consider a scenario where gcd(a, m) > 1. Take the congruence 6x 9 (mod 12). Here, gcd(6, 12) = 6, and since 6 divides 9, solutions exist. We can simplify the congruence by dividing by the greatest common divisor to obtain x 3/2 (mod 2), which is equivalent to x 1 (mod 2). This means x can be any odd number. In general, if solutions exist when gcd(a,m) > 1, we would obtain a simpler congruence where gcd(a',m')=1 where a' = a/gcd(a,m) and m' = m/gcd(a,m). This is the key to unraveling complex congruences.

The solutions to a linear congruence are not just isolated numbers; they form a set of integers which can be represented by a single solution plus multiples of a certain integer. This pattern is directly related to the concept of modular arithmetic's cyclic nature. The number of solutions, as previously mentioned, is directly related to the greatest common divisor of a and m. Comprehending this relationship lays a fundamental groundwork for approaching more advanced problems in number theory, such as the Chinese Remainder Theorem which we will encounter shortly. This theorem deals with systems of congruences and demonstrates how solving linear congruences individually can be combined to solve a larger, more complex system.

Understanding linear congruences opens doors to more advanced concepts. It's a building block—a seemingly simple tool, yet its power is evident in its wide-reaching impact. From simple examples to complex systems, the principles we've discussed here provide the essential framework for tackling more challenging problems in number theory. The solutions themselves, and the methods employed to reach them, highlight the inherent elegance and precision at the heart of mathematical theory.

3.3 Chinese Remainder Theorem

The Chinese Remainder Theorem (CRT) offers a powerful method for solving systems of simultaneous congruences. Imagine you have a collection of clues, each revealing a piece of information about an unknown number modulo different integers. The CRT provides a systematic way to combine these seemingly disparate clues to determine the unique solution, or at least find a family of solutions. This theorem finds applications far beyond pure mathematics, showing up in areas like cryptography and computer science, where its ability to handle multiple constraints simultaneously proves invaluable.

Consider a simple example: Suppose we know a number leaves a remainder of 2 when divided by 3, and a remainder of 3 when divided by 5. In the language of congruences, we can write this as x 2 (mod 3) and x 3 (mod

5). The CRT provides a structured approach to find the integer x satisfying both congruences. The beauty of the CRT lies in its ability to handle systems with more than two congruences, a feat that wouldn't be easily tackled using basic modular arithmetic alone.

The theorem's power lies in its constructive proof, which not only establishes existence but also provides an explicit method for finding the solution. This constructive nature allows for direct application in various computational scenarios, unlike some existence theorems that only assert the solution's existence without offering practical guidance on how to find it. The algorithm involves calculating the least common multiple of the moduli (the numbers we're taking remainders modulo), employing the extended Euclidean algorithm to find multiplicative inverses, and carefully combining the results. This combination process is more than just simple addition; it's a weighted sum, reflecting the importance of each congruence within the overall solution.

However, the CRT is not without its limitations. The theorem requires that the moduli in the system of congruences are pairwise relatively prime—that is, no two moduli share a common divisor greater than 1. If this condition is not met, the system might have no solution or multiple solutions, complicating the analysis. Detecting this condition is crucial before attempting to apply the CRT; otherwise, one might invest effort into a procedure that yields inaccurate or meaningless results. Recognizing this limitation is crucial for correctly applying the theorem and interpreting its output.

Let's delve into the specifics of the algorithm. We start by finding the product N of all the moduli. For each modulus, we calculate its complementary product N, which is N divided by that modulus. Then, we use the extended Euclidean algorithm—a process for finding the greatest common divisor of two integers and representing it as a linear combination of those integers—to find the multiplicative inverse of N modulo the corresponding modulus. This inverse allows us to "weight" the remainders appropriately, ensuring that the final combined solution satisfies all the congruences simultaneously. It's a clever use of modular inverses to effectively isolate and combine the information from each individual congruence.

The CRT's elegance lies in its ability to unify seemingly unrelated pieces of information. Consider a scenario involving several independent observations, each providing a partial glimpse into an unknown quantity. The CRT allows us to assemble these fragments into a coherent whole, revealing a previously obscured truth. This unifying power extends to practical applications in areas such as scheduling, cryptography, and even solving certain types of puzzles that involve modular constraints. The theorem's ability to connect disparate pieces of information makes it a fundamental tool in solving complex problems.

Furthermore, the Chinese Remainder Theorem provides insights into the structure of modular arithmetic. It highlights the interplay between different moduli and demonstrates how solutions can be constructed by carefully balancing the information provided by each congruence. This perspective is valuable for a deeper understanding of modular arithmetic and its applications. It's more than just a problem-solving technique; it's a window into the underlying mathematical structure.

The efficient implementation of the CRT is a significant aspect of its practical applicability. Optimized algorithms, leveraging fast modular arithmetic techniques, ensure that even systems with numerous congruences can be solved efficiently. This efficiency is particularly crucial in computationally intensive applications, such as cryptography, where speed is paramount. High-performance implementations of the CRT are essential components in many cryptographic systems, ensuring their speed and security.

In conclusion, the Chinese Remainder Theorem stands as a testament to the elegance and power of number theory. Its ability to solve systems of simultaneous congruences extends far beyond theoretical considerations, providing a valuable tool for addressing real-world problems across various disciplines. From cryptography to scheduling, the theorem's impact is widespread, highlighting its significance in both pure and applied mathematics. Mastering the CRT is not just about learning a theorem; it is about gaining a deeper understanding of modular arithmetic and its far-reaching implications.

3.4 Fermat's Little Theorem

Fermat's Little Theorem offers a powerful tool for understanding the behavior of integers modulo a prime number. It states that if p is a prime number, then for any integer a, the number ap \item a is an integer multiple of p. This seemingly simple statement has profound implications across number theory, influencing areas like cryptography and primality testing. We can express this concisely as a ap (mod p).

This theorem isn't just a theoretical curiosity; it provides a practical method for simplifying calculations within modular arithmetic. For example, consider finding the remainder when 2100 is divided by 7. Direct calculation is impractical, but Fermat's Little Theorem provides a shortcut. Since 7 is prime, we know that 27 2 (mod 7). We can then use properties of modular exponentiation to efficiently determine the remainder. Repeated applications of the theorem, cleverly exploiting modular arithmetic's cyclic nature, allow us to tackle exponentially large exponents with relative ease.

The proof itself is elegant and insightful, relying on properties of prime numbers and binomial coefficients. One common approach utilizes the combinatorial identity (a + b)p, expanded via the binomial theorem and considering the congruence modulo p. The binomial coefficients, except for the first and last, are divisible by p for prime p, leading to a simplification directly related to Fermat's Little Theorem's statement. Understanding this proof illuminates the deep connection between combinatorics and number theory.

It's crucial to note that the converse of Fermat's Little Theorem doesn't hold. If an a (mod n) for some integer a, it doesn't necessarily mean n is prime. Composite numbers that satisfy this congruence for all integers a are called Carmichael numbers. These numbers are rare and their existence complicates primality testing, highlighting the need for more sophisticated primality tests in practical applications. This subtlety underscores the importance of a thorough understanding of the theorem's limitations, as much as its capabilities.

Beyond its use in simplification, Fermat's Little Theorem forms a critical foundation for numerous advanced concepts. It serves as a stepping stone to Euler's totient theorem, a generalization applicable to composite numbers. It plays a significant role in cryptographic algorithms like RSA encryption, where the security rests on the computational difficulty of factoring large numbers. The algorithm cleverly exploits the properties of modular exponentiation, ensuring the message's secrecy hinges on the practical inability to perform the equivalent of computing large exponents modulo composite numbers. Consequently, mastering Fermat's Little Theorem isn't merely academic; it's key to understanding a substantial portion of modern cryptography.

Let's consider a concrete example of its application. Imagine you need to determine if a large number, say N, is a possible prime. Instead of trying direct factorization (a computationally expensive process for large numbers), Fermat's Little Theorem suggests a probable primality test. Select a random integer a and calculate aN-1 (mod N). If the result is not congruent to 1, you know that N is certainly composite. However, if it is 1, N might be prime—but there's a small chance it could be a Carmichael number, leading to a false positive. This test is not foolproof, yet its efficiency makes it a valuable tool in preliminary primality screening before resorting to more rigorous primality tests. This illustrates that even though not infallible, Fermat's Little Theorem provides a practical, probabilistic way to approach primality tests.

The seemingly simple elegance of Fermat's Little Theorem belies its deep significance in number theory and its practical applications. Its proof reveals a beautiful connection between combinatorial identities and modular arithmetic. Its limitations, such as the existence of Carmichael numbers, highlight the need for a nuanced understanding of its strengths and weaknesses. From simplifying complex modular calculations to forming the cornerstone of cryptographic algorithms, its impact is far-reaching, demonstrating the power of seemingly simple mathematical concepts to solve complex problems. Understanding and applying Fermat's Little Theorem is a fundamental step for anyone venturing further into the fascinating world of number theory. This theorem serves as a bridge, linking basic number theory concepts to advanced fields, making it an indispensable tool in the mathematician's arsenal. Its pervasive influence underscores the interconnected nature of mathematical fields, demonstrating how seemingly disparate branches are connected through fundamental theorems. Its legacy extends far beyond the mathematical realm, enriching areas such as cryptography and computer science, proving its enduring relevance in both theoretical and applied mathematics.

3.5 Euler's Totient Theorem

Euler's totient function, denoted (n), counts the positive integers less than or equal to n that are relatively prime to n. This means they share no common factors other than 1. For instance, (10) = 4 because 1, 3, 7, and 9 are relatively prime to 10. Understanding this function is key to grasping Euler's Totient Theorem, a powerful tool in number theory with significant applications in cryptography.

Calculating (n) directly can be tedious for large n. However, if we know the prime factorization of n, we can utilize a formula for efficient computation. If n = p^k \emph{ p^k } ... \emph{ p^k is the prime factorization of n, where p are distinct primes and k are positive integers, then (n) = n } (1 \item 1/p) \emph{ (1 - 1/p) } ... \emph{ (1 - 1/p). This formula simplifies the calculation considerably, allowing us to determine (n) without directly counting relatively prime integers. For example, to find (12), we factor 12 as 2^2 } 3. Then, (12) = 12 \emph{ (1 - 1/2) } (1 - 1/3) = 12 \emph{ (1/2) } (2/3) = 4, matching our earlier direct count.

Euler's Totient Theorem states that if a and n are relatively prime, then a^(n) 1 (mod n). This congruence means that a raised to the power of (n) leaves a remainder of 1 when divided by n. This seemingly simple statement has profound implications. Its proof involves considering the set of integers from 1 to n that are relatively prime to n. Multiplying each element in this set by a (mod n) results in a permutation of the same set, a fact crucial for demonstrating the theorem. This elegant proof highlights

the intricate connections within number theory.
Let's consider an example: Let's take a = 3 and n = 10. We already know (10) = 4. According to Euler's theorem, 3 1 (mod 10). Indeed, 3 = 81, and 81 leaves a remainder of 1 when divided by 10. This confirms the theorem's validity for this specific case. However, the true power of this theorem lies in its generality; it holds for any pair of relatively prime integers a and n.
The practical significance of Euler's Totient Theorem is undeniable. It forms the foundation of RSA cryptography, a widely used public-key cryptosystem. RSA relies on the difficulty of factoring large numbers. The security of RSA hinges on the fact that while it's easy to compute a^(n) (mod n) given a, n, and (n), it's computationally infeasible to determine (n) from n alone if n is a product of two large prime numbers. This asymmetry is what allows for secure communication. Specifically, the encryption and decryption processes involve modular exponentiation using the totient function. Furthermore, Euler's theorem provides a powerful tool for solving certain types of congruences. If we want to solve a congruence of the form x^k b (mod n), where k and (n) are relatively prime, then Euler's theorem can often be used to simplify the problem. This involves manipulating the congruence using the properties of modular arithmetic and applying Euler's theorem to reduce the exponent. The practical application of this in various fields of mathematics and computer science is extensive, offering many avenues for further investigation and research.
Finally, Euler's Totient Theorem underscores the elegance and depth of number theory. Its seemingly simple statement hides a profound math-

ematical truth with far-reaching consequences. The connection between the seemingly abstract concept of relatively prime numbers and the practical application in cryptography showcases the power and beauty of mathematical discovery. It serves as a reminder that even seemingly esoteric mathematical concepts can have profound real-world applications. Mastering this theorem opens doors to a deeper understanding of advanced number theory and its significant impact on modern technology. The theorem's elegance and power make it a cornerstone of modern mathematics and a crucial concept for any aspiring number theorist.

4 Diophantine Equations

Linear Diophantine equations, those elegant puzzles involving only integers, hold a special place in number theory. Their simple form—ax + by = c, where a, b, and c are integers—belies a depth of mathematical insight. Finding integer solutions, or proving their absence, relies on understanding the greatest common divisor (GCD) of a and b. Specifically, a solution exists if and only if the GCD of a and b divides c. This fundamental theorem provides a powerful tool for analysis, allowing us to quickly determine solvability before embarking on more complex calculations. Furthermore, using the Euclidean algorithm, we can efficiently compute the GCD and, if a solution exists, express it in terms of the coefficients a and b. This allows us to generate all possible solutions, revealing the infinite family of solutions that often characterize these equations.

The Euclidean Algorithm itself is a marvel of efficiency, a testament to the power of iterative processes. It systematically reduces the problem of finding the GCD of two numbers to a sequence of smaller, equivalent problems, ultimately arriving at the greatest common divisor. The process not only determines the GCD, but also provides coefficients x and y that satisfy the equation $ax + by = gcd(a, b)$. This is crucial because these coefficients form the basis for constructing all integer solutions to the more general equation $ax + by = c$. Mastering the Euclidean algorithm is paramount; it's the key that unlocks the door to solving linear Diophantine equations. Practice is essential. Work through various examples; experiment with different coefficients. This will solidify your understanding and allow you to tackle more challenging problems with confidence.

Pythagorean triples, those harmonious sets of integers (x, y, z) satisfying the equation $x^2 + y^2 = z^2$, possess a captivating geometric interpretation. They represent the side lengths of right-angled triangles with integer sides. The simplest and most famous example, (3, 4, 5), has captivated mathematicians for millennia. But how do we find others? Euclid's formula provides an elegant and systematic method for generating these triples. It involves two arbitrary integers, m and n, which, when correctly applied, yield the sides of a Pythagorean triple. However, not all triples are generated by Euclid's formula; some triples arise from unexpected sources, leading to intriguing questions regarding their classification and distribution. The study of Pythagorean triples reveals a connection between seemingly disparate areas of mathematics – algebra, geometry and number theory. This demonstrates the interconnectedness of mathematical concepts and encourages an exploration beyond simple memorization.

Beyond these basic forms, Diophantine equations extend to higher degrees and greater complexity. For example, Fermat's Last Theorem, a problem that haunted mathematicians for centuries, deals with the equation x +

$y = z$ for integer values of n greater than 2. The proof, finally achieved by Andrew Wiles, involved incredibly advanced mathematical techniques, showcasing the profound depth and challenges hidden within Diophantine equations. While the full proof is beyond the scope of an introductory text, understanding the simpler cases lays a crucial foundation for appreciating the complexity of the problem. Remember, solving these equations is about more than just finding answers. It involves developing problem-solving strategies, learning to think critically and systematically, and applying the tools of number theory with precision.

The study of Diophantine equations is a journey into the heart of number theory, a realm where abstract concepts meet practical applications. From the seemingly simple linear equations to the profound challenges posed by Fermat's Last Theorem, these equations offer a rich field of exploration, testing mathematical intuition and problem-solving skills. Remember, the path to mastery lies in practice. Explore examples, work through problems, and actively seek to understand the underlying mathematical principles. The rewards of this journey are immense: a deeper appreciation for the elegance and power of number theory and the satisfaction of solving these ancient mathematical puzzles. Don't be afraid to explore further. Seek out additional resources, challenge yourself with more complex problems and ultimately, embrace the beauty and complexity inherent in the study of Diophantine equations. The journey is as rewarding as the destination.

4.1 Linear Diophantine Equations

A linear Diophantine equation is an equation of the form $ax + by = c$, where a, b, and c are integers, and we seek integer solutions for x and y. The existence of solutions hinges on the greatest common divisor (GCD) of a and b. Specifically, if the GCD of a and b does not divide c, then no integer solutions exist. This is a fundamental theorem, easily proven through consideration of modular arithmetic and Bézout's identity. Understanding this criterion is your first step to mastering this area.

If, however, the GCD of a and b does divide c, then integer solutions exist. Finding them requires a slightly more involved process, typically employing the extended Euclidean algorithm. This algorithm provides not only the GCD but also integers x and y satisfying the equation $ax + by = GCD(a, b)$. From this particular solution, we can generate infinitely many others; understanding this family of solutions is critical.

Let's consider an example. Suppose we have the equation $6x + 15y = 9$. The $GCD(6, 15) = 3$, and 3 divides 9. Employing the extended Euclidean algorithm (or simply by observation), we find that $6(-2) + 15(1) = 3$. Multiplying this equation by 3, we obtain $6(-6) + 15(3) = 9$. Thus, $x = -6$ and $y = 3$ provide a particular solution to our original equation.

The general solution is then given by $x = x + (b/GCD(a, b))n$ and $y = y$ \item (a/GCD(a, b))n, where n is any integer. In our example, the general solution becomes $x = -6 + 5n$ and $y = 3 - 2n$. By varying n, one can generate an infinite number of integer solutions (e.g., $n = 0$ gives $x = -6$, $y = 3$; $n = 1$ gives $x = -1$, y

= 1; n = 2 gives x = 4, y = -1, and so on).

The importance of linear Diophantine equations extends far beyond theoretical number theory. They frequently appear in various applied mathematical contexts. Consider problems involving resource allocation, where 'a' represents the units of resource A, 'b' represents units of resource B, and 'c' represents the total required resources. Finding integer solutions (x and y) determines the quantities of A and B to use optimally.

Moreover, linear Diophantine equations serve as a crucial building block for understanding more complex Diophantine equations. Many advanced techniques used in tackling nonlinear equations are grounded in the principles and methods of solving linear ones. Mastering this seemingly simple equation type empowers you to approach more challenging Diophantine problems effectively.

Consider further the case where a and b are coprime, that is, GCD(a, b) = 1. In this simplified scenario, the equation ax + by = c always has integer solutions if c is an integer. The process of finding these solutions remains similar: the extended Euclidean algorithm delivers a particular solution, and a general solution is then easily derived through the formula mentioned earlier. The beauty of this simplified case lies in its clarity and immediate applicability.

Imagine, for example, a scenario in a manufacturing plant. You have two types of machines, producing components at rates 'a' and 'b' per hour respectively. Your daily production goal is 'c' components. Solving the linear Diophantine equation ax + by = c, where x and y represent the operating hours of the two machines, directly determines the optimal production schedule. This practical application highlights the immediate usefulness of understanding these seemingly abstract equations.

However, it's equally crucial to remember the cases where no solution exists. It's not enough to blindly apply the algorithm; the initial check for divisibility is essential. Failing to confirm the divisibility of c by GCD(a, b) can lead to fruitless attempts at finding a solution, highlighting the need for a thorough understanding of the fundamental theorem that governs their solvability. Therefore, always start by assessing the GCD of your coefficients.

This thorough examination of linear Diophantine equations lays the groundwork for your exploration into higher-level Diophantine equations. It underscores that often, the simplicity of the fundamental concepts belies their profound implications and wide applicability. A solid grasp of this topic is essential for further advancement in number theory. The methods used here—the Euclidean algorithm, the understanding of modular arithmetic, and the careful analysis of the general solution—will serve as invaluable tools as you progress.

4.2 Pythagorean Triples

A Pythagorean triple consists of three positive integers a, b, and c, satisfying the equation $a^2 + b^2 = c^2$. These triples represent the lengths of the sides of a right-angled triangle, a fundamental concept in geometry. The simplest and most well-known example is the triple (3, 4, 5), since $3^2 + 4^2$

= 9 + 16 = 25 = 5^2. Understanding Pythagorean triples involves more than just recognizing a few examples; it delves into the underlying mathematical structures and allows us to generate infinitely many such triples.

One approach to finding Pythagorean triples is using Euclid's formula, a powerful tool derived from simple algebraic manipulation. This formula states that for any two positive integers m and n, where m > n, the triple (m^2 \item n^2, 2mn, $m^2 + n^2$) will always satisfy the Pythagorean equation. For example, if we let m = 2 and n = 1, we get (3, 4, 5), confirming the formula's validity. Changing the values of m and n generates a vast array of new triples. The elegance of Euclid's formula lies in its ability to systematically generate all primitive Pythagorean triples—those triples where a, b, and c share no common divisor greater than 1. Exploring different combinations of m and n will quickly demonstrate the formula's effectiveness.

The concept of primitive triples is crucial because any non-primitive triple is simply a multiple of a primitive one. For example, (6, 8, 10) is a multiple of (3, 4, 5). Focusing on primitive triples allows us to concentrate on the fundamental building blocks of all Pythagorean triples, simplifying the analysis significantly. This focus on primitive triples helps to avoid redundancy and provides a more streamlined understanding of the overall structure. The systematic generation of these fundamental triples, therefore, is a key aspect of understanding the entire family of Pythagorean triples.

Beyond Euclid's formula, we can explore the geometric interpretation of Pythagorean triples. Visualizing these triples as sides of right-angled triangles provides a geometric understanding of the relationship between the integers. Consider constructing a square with side length 'a' and another square with side length 'b'. The area of the combined squares will equal the area of a third square with side length 'c'. This geometric representation visually demonstrates the essence of the Pythagorean theorem and provides an intuitive understanding of the concept that goes beyond the abstract algebraic representation. This visualization makes the concept more accessible and memorable.

The study of Pythagorean triples extends to more advanced areas of number theory. For instance, Fermat's Last Theorem, a renowned problem that took centuries to solve, is closely related. This theorem states that there are no integer solutions to the equation $x + y = z$ for any integer value of n greater than 2. The Pythagorean equation (n=2) is the only case where such integer solutions exist, highlighting the unique nature of Pythagorean triples within the broader context of Diophantine equations. Understanding Pythagorean triples therefore provides a foundational understanding that allows for a deeper appreciation of this more complex concept.

Furthermore, Pythagorean triples have applications beyond theoretical mathematics. They find uses in various fields, including computer graphics and cryptography. In computer graphics, understanding Pythagorean triples allows for efficient calculations related to right-angled triangles, optimizing processes involved in image rendering and geometric modelling. In cryptography, the inherent mathematical structure of these triples can be leveraged for various security protocols. The ability to efficiently generate and manipulate these triples is therefore not just a mathematical curiosity but holds practical importance across multiple fields. In summary, the exploration of Pythagorean triples offers a fascinating journey through number theory. From Euclid's formula, which allows for the systematic generation of these triples, to their geometric interpretation and their connection to more advanced number-theoretic concepts like Fermat's Last Theorem, the study of Pythagorean triples is rich and rewarding. The practical applications further highlight their importance beyond the purely theoretical realm, reinforcing their significance in the wider mathematical landscape. Understanding these triples thoroughly provides a solid foundation for further explorations within number theory and related disciplines. The seemingly simple equation hides a complexity and elegance that continue to fascinate mathematicians to this day. This chapter only scratches the surface; dedicated study will reveal even more of the fascinating intricacies of this fundamental mathematical concept.

5 Advanced Congruence Topics

We've explored congruences; now, let's delve into the fascinating world of quadratic residues. A quadratic residue modulo n is an integer a such that there exists an integer x satisfying the congruence x^2 a (mod n). Think of it like this: can you find a perfect square that leaves a remainder of a when divided by n? If so, a is a quadratic residue. If not, it's a quadratic non-residue. This seemingly simple concept opens doors to many intricate mathematical problems. For instance, consider the congruence x^2 2 (mod 7). We can test values of x: 1^2 1 (mod 7), 2^2 4 (mod 7), 3^2 2 (mod 7). Thus, 2 is a quadratic residue modulo 7. However, if we consider x^2 3 (mod 7), there is no solution, making 3 a quadratic non-residue. Determining quadratic residuosity becomes significantly more complex as n increases and introduces prime factorization's importance.

The Legendre symbol provides a concise way to determine quadratic residuosity when the modulus is an odd prime. It's denoted as (a/p), where a is an integer and p is an odd prime. The symbol evaluates to 1 if a is a quadratic residue modulo p, -1 if a is a quadratic non-residue, and 0 if p divides a. The beauty of the Legendre symbol lies in its elegant calculation, often employing properties like multiplicativity: (ab/p) = (a/p)(b/p). For example, let's determine if 5 is a quadratic residue modulo 11. Using the Legendre symbol, we have (5/11). Calculations, employing properties of the symbol and perhaps Euler's criterion (a powerful tool for this context), allow efficient computation. We don't need to exhaustively test all possible squares. This is invaluable in cryptography, as the efficiency of the calculations heavily influences the speed of encryption and decryption.

Building upon the Legendre symbol, we encounter the Jacobi symbol. This extends the concept to composite moduli, providing a generalization beyond odd primes. The Jacobi symbol (a/n) is defined for an integer a and an odd positive integer n by factoring n into its prime components as n = p1e1 p2e2...pkek, we have (a/n) = (a/p1)e1(a/p2)e2...(a/pk)ek. Notice how it's a product of Legendre symbols! While the Jacobi symbol doesn't directly determine quadratic residuosity for composite moduli—a non-trivial distinction—its properties are extremely useful. It's computationally efficient and shares some crucial properties with the Legendre symbol, such as multiplicativity. These properties are instrumental in various algorithms, including primality testing and cryptographic applications. The fact that the Jacobi symbol is readily computable even for large numbers makes it a powerful tool in more advanced applications such as the Solovay-Strassen primality test.

Let's illustrate. Suppose we want to determine (10/15). We first factor 15 as 3 × 5. The Jacobi symbol (10/15) becomes (10/3)(10/5). We then evaluate the individual Legendre symbols. (10/3) = (1/3) = 1, since 1^2 1 (mod 3). (10/5) =

0 because 5 divides 10. Therefore, (10/15) = 1 * 0 = 0. This demonstrates that the Jacobi symbol effectively handles composite moduli, and its zero value indicates that 10 is divisible by a prime factor of 15. The computational efficiency of the Jacobi symbol is crucial for advanced cryptographic schemes. Now, let's connect these advanced congruence topics back to our prior discussions. Remember linear congruences? Solving ax b (mod n) is fundamental. Quadratic congruences, of the form x^2 a (mod n), introduce a new layer of complexity. The Legendre and Jacobi symbols provide powerful tools to analyze the solvability of such congruences. They don't provide direct solutions, but they tell us whether solutions exist—a critical first step. This informs strategies for finding actual solutions, a significant advancement over a brute-force approach, especially when n is large.
Furthermore, recall the Chinese Remainder Theorem. It helps solve systems of linear congruences. Although we don't directly use quadratic residues in the theorem itself, the concepts are intertwined. When considering systems involving quadratic congruences, the tools we've discussed become invaluable for analyzing the existence and number of solutions, making the analysis far more systematic than trial and error.
Finally, the implications for cryptography are immense. The difficulty of determining quadratic residuosity for large moduli forms the basis of some cryptographic schemes. These schemes cleverly exploit the computational difference between determining whether a number is a quadratic residue and actually finding a square root modulo a large number. This asymmetry forms the bedrock of security in many modern cryptographic

systems, highlighting the profound practical significance of our theoretical discussions. Understanding quadratic residues and the associated symbols is not just a theoretical exercise, it is a key to understanding the security behind many online transactions. Our exploration of these advanced congruence topics has thus revealed not only the mathematical elegance but also the practical power at the heart of number theory.

5.1 Quadratic Residues

We'll explore quadratic residues, a concept central to number theory. Imagine you're working modulo a prime number p. A quadratic residue is an integer a such that there exists an integer x satisfying the congruence x^2 a (mod p). Think of it like this: can you find a square root of a within the modular arithmetic system? If you can, a is a quadratic residue; otherwise, it's a quadratic non-residue. This seemingly simple definition opens a world of fascinating mathematical properties and applications.

The determination of whether a number is a quadratic residue is not always straightforward. It isn't just a matter of trying every possible square root. For small primes, brute force might work, but for larger primes, more sophisticated tools are needed. This is where the power of Legendre symbol comes into play. This symbol, denoted as (a/p), provides a concise way to express whether 'a' is a quadratic residue modulo 'p'. It elegantly encapsulates the solution, either +1 if a is a quadratic residue, -1 if it's a non-residue, and 0 if p divides a.

Understanding the Legendre symbol is crucial because it allows us to move beyond simple trial-and-error. It transforms a potentially complex calculation into a clear, compact representation. For instance, knowing (2/p) allows us to quickly assess if 2 is a quadratic residue modulo any given prime p without laborious checking. This efficiency is particularly valuable when dealing with larger numbers. The calculation of the Legendre symbol often involves the use of properties such as quadratic reciprocity, a deep and elegant theorem connecting the quadratic character of two different primes.

Let's consider a concrete example. Take the prime p = 7. Let's check if 2 is a quadratic residue modulo 7. We examine the squares modulo 7: 1^2 1, 2^2 4, 3^2 2, 4^2 2, 5^2 4, 6^2 1 (mod 7). Note that 2 appears as a quadratic residue. Thus, (2/7) = 1. On the other hand, if we consider 3, it is a quadratic non-residue modulo 7 because it doesn't show up amongst the squares modulo 7, thus (3/7) = -1. This simple example illustrates the fundamental idea of quadratic residues.

But what happens when we move beyond prime moduli? The Jacobi symbol extends the idea of the Legendre symbol to composite odd numbers. The Jacobi symbol (a/n), where n is a composite odd number, is a generalization of the Legendre symbol. It's calculated by breaking down n into its prime factors, computing Legendre symbols for each prime factor, and then multiplying the results. Though the Jacobi symbol doesn't directly tell us whether a is a quadratic residue modulo n, it retains some important properties that make it useful in various algorithms. For instance, if (a/n) = -1, we know immediately that 'a' is not a quadratic residue modulo 'n'.

The applications of quadratic residues and their related symbols extend far beyond theoretical number theory. They form the foundation for several cryptographic algorithms. The strength of algorithms like the quadratic sieve, used for factoring large integers, relies heavily on the properties of quadratic residues. Understanding the distribution and properties of quadratic residues is essential for assessing the security of these widely used cryptographic systems. Furthermore, the efficiency of calculations related to quadratic residues is paramount in the design and implementation of robust and efficient cryptographic techniques.

Consider the computational aspects involved. While determining quadratic residuosity for smaller numbers might be easily handled through direct calculation, the computational cost grows exponentially with the size of the modulus. For extremely large numbers, efficient algorithms are crucial. This is where understanding the properties of the Legendre and Jacobi symbols becomes even more relevant. Sophisticated techniques and algorithms exploit these properties to reduce the computational burden significantly. It's a testament to the elegance and power of mathematical theory: a seemingly abstract concept allows for practical advancements in applied fields like cryptography.

In conclusion, quadratic residues are a critical component of number theory. Their deceptively simple definition gives rise to profound results and applications. The Legendre and Jacobi symbols provide elegant tools to work with quadratic residues, allowing for efficient computations. This understanding is essential not only for a deep grasp of number theory but also for developing and analyzing the security of modern cryptographic systems. The interplay between abstract theory and practical application showcases the power and beauty of mathematics. The continued exploration of these concepts remains at the forefront of ongoing research in both pure and applied mathematics, highlighting their enduring importance in the field. Further study will reveal even more depth and connection to other fascinating areas of mathematics.

5.2 Legendre Symbol

The Legendre symbol, denoted as (a/p), is a powerful tool for determining whether an integer 'a' is a quadratic residue modulo a prime number 'p'. It provides a concise way to answer the question: Does there exist an integer x such that x^2 a (mod p)? This seemingly simple question has profound implications across number theory, especially in the study of quadratic congruences and their applications. Understanding the Legendre symbol is crucial for mastering more advanced topics, like the Jacobi symbol and various cryptographic techniques.

The symbol itself is defined as +1 if 'a' is a quadratic residue modulo 'p', -1 if 'a' is a quadratic non-residue, and 0 if 'p' divides 'a'. This compact notation provides immediate insight into the solvability of a specific type of congruence. For instance, if (a/p) = 1, we know a solution exists; if (a/p) = -1, no solution exists. The case where (a/p) = 0 is trivial, as 'a' is clearly congruent to 0 modulo 'p'. Remember that 'p' must always be an odd prime for the

Legendre symbol to be properly defined.

Calculating the Legendre symbol directly involves examining all possible values of x^2 modulo 'p' to see if any are congruent to 'a'. This method, while conceptually straightforward, becomes computationally expensive for large values of 'p'. Thankfully, we have the aid of Euler's criterion which significantly streamlines the calculation. Euler's criterion states that (a/p) a^((p-1)/2) (mod p). This allows us to compute the Legendre symbol using modular exponentiation, a significantly faster method than direct enumeration, particularly for large prime numbers.

Let's consider a concrete example. Suppose we want to determine if 7 is a quadratic residue modulo 11. Using Euler's criterion, we calculate 7^((11-1)/2) 7^5 (mod 11). This simplifies to 7^5 16807 16807 \item 1527(11) 4 (mod 11). This is not congruent to 1, meaning 7 is not a quadratic residue modulo 11; thus, (7/11) = -1. This calculation demonstrates the elegance and efficiency of Euler's criterion in determining the Legendre symbol's value, a significant advantage over brute-force methods.

Further enhancing the utility of the Legendre symbol is the law of quadratic reciprocity. This remarkable theorem connects the Legendre symbol (a/p) with (p/a), where 'a' is also an odd prime. Specifically, it states that (a/p) = (p/a) if both 'a' and 'p' are congruent to 1 modulo 4, or if both are congruent to 3 modulo 4. However, if one is congruent to 1 modulo 4 and the other to 3 modulo 4, then (a/p) = -(p/a). This reciprocity allows for a recursive evaluation of Legendre symbols, significantly reducing computational complexity, particularly when dealing with larger numbers.

The law of quadratic reciprocity, in conjunction with Euler's criterion and properties of the Legendre symbol such as (ab/p) = (a/p)(b/p), provides a systematic and efficient way to compute (a/p). For instance, when computing (17/23), we can utilize the reciprocity law and the supplementary rules to break the calculation down into smaller, more manageable Legendre symbols, significantly simplifying the entire process. This systematic approach makes the Legendre symbol computationally feasible even with considerably larger numbers.

The practical applications of the Legendre symbol extend far beyond theoretical calculations. Its impact is particularly evident in cryptography. Many cryptographic algorithms rely on the difficulty of determining quadratic residues modulo large primes. Understanding the Legendre symbol's properties and its efficient calculation is thus essential for comprehending and developing secure cryptographic systems. The Legendre symbol represents a cornerstone in the theoretical framework of these systems, establishing a direct link between abstract mathematical concepts and practical security applications. The insights offered by the Legendre symbol are not confined to the theoretical realm; they provide invaluable tools for designing and analyzing crucial cryptographic techniques.

Moreover, the Legendre symbol is fundamental to the study of higher-order reciprocity laws and deeper investigations within number theory. Its connection to Gauss sums, its role in characterizing quadratic residues, and its generalizations to the Jacobi symbol all underscore its significance as a cornerstone within the field. Understanding this symbol opens doors to various areas of number theory, offering insights into sophisticated concepts such as the distribution of quadratic residues and their profound consequences. This basic yet powerful tool illuminates some of the most interesting complexities of number theory.

In summary, the Legendre symbol is a fundamental concept in number theory with practical applications in cryptography and deeper theoretical connections throughout the field. Its concise definition, efficient calculation via Euler's criterion, and the law of quadratic reciprocity make it an essential tool for anyone serious about mastering the concepts of quadratic residues and the intricate world of number theory. A firm grasp of this symbol will serve as a solid foundation for exploring more advanced topics, reinforcing the interconnectedness of various areas within this rich and complex subject.

5.3 Jacobi Symbol

The Jacobi symbol, a powerful generalization of the Legendre symbol, provides a convenient tool for determining quadratic residues modulo composite numbers. Unlike the Legendre symbol, which is only defined for prime moduli, the Jacobi symbol extends its functionality to composite odd integers. This extension is crucial because many applications in number theory and cryptography involve composite numbers.

Its calculation relies on the prime factorization of the modulus, a process which itself reveals important structural details about the number in question. Each prime factor contributes its own Legendre symbol to the overall Jacobi symbol calculation, a multiplicative process that reveals unexpected patterns and relationships. We break down a composite odd integer n into its prime factorization, n = p1e1p2e2...pkek, where each pi is a distinct prime number and ei its corresponding exponent. For any integer a, the Jacobi symbol (a/n) is defined as the product of the Legendre symbols (a/pi)ei for each prime factor pi.

This seemingly straightforward definition hides a surprising depth. Consider the implications: The Jacobi symbol's calculation involves a sequence of computations on individual primes, which allows the application of known properties of the Legendre symbol. Importantly, the Jacobi symbol (a/n) can be equal to 1 even when a is not a quadratic residue modulo n. This contrasts with the Legendre symbol, where (a/p) = 1 only if a is a quadratic residue modulo the prime p. This crucial difference highlights the distinctions and expands the potential applications.

One key application of the Jacobi symbol lies in the realm of primality testing. Algorithms such as the Solovay-Strassen primality test efficiently utilize the Jacobi symbol to probabilistically determine whether a given number is prime. The speed and efficiency of these algorithms hinge on the ease of computation of the Jacobi symbol, significantly reducing computational overhead compared to direct factorization methods.

The computational advantage stems from its recursive nature. The Jacobi symbol can be computed efficiently using the quadratic reciprocity law, a fundamental theorem that connects the Legendre symbols of two primes. This reciprocity enables us to recursively reduce the calculation of the Jacobi symbol to smaller components, simplifying the process and offering computational shortcuts not available with direct factorization.

Furthermore, the Jacobi symbol finds applications in cryptography, particularly in probabilistic primality testing. Its efficient computation significantly speeds up the generation of large prime numbers which form the basis of many widely used cryptographic systems. These prime numbers are essential in securing online transactions and ensuring data privacy, highlighting the crucial role of this apparently abstract mathematical concept.

Let's explore a concrete example. Suppose we want to compute the Jacobi symbol (10/15). First, we factorize 15 as 3×5. This allows us to express (10/15) as (10/3)(10/5). Now, we apply the Legendre symbol definition. (10/3) is equivalent to (1/3) which equals 1, because 1 is a quadratic residue modulo 3. Similarly, (10/5) is equivalent to (0/5), which is 0 by definition. Therefore, $(10/15) = 1 \times 0 = 0$. This demonstrates the straightforward calculation of the Jacobi symbol using prime factorization and the properties of the Legendre symbol.

In contrast, calculating directly whether 10 is a quadratic residue modulo 15 requires a different, more laborious approach. The Jacobi symbol offers an elegant workaround, reducing the complexity of this specific determination. This computational efficiency becomes exponentially more significant as the numbers involved grow larger, making it a vital tool in modern computational number theory. This method illustrates the elegance and efficiency of the Jacobi symbol's computational aspects in solving problems involving quadratic residues, far exceeding the capabilities of direct approaches.

The power of the Jacobi symbol lies in its ability to bridge the gap between theory and computation. It provides a practical algorithm for determining quadratic residuosity, a problem of fundamental importance in number theory. Its applications extend beyond the theoretical realm, deeply influencing the development and implementation of cryptographic systems and efficiency within primality testing. The Jacobi symbol is not merely a theoretical curiosity; it is a powerful computational tool that has profound implications in modern technology. The beauty of mathematics often lies in such unexpected connections, transforming abstract concepts into practical instruments.

6 Applications in Cryptography

The RSA algorithm, a cornerstone of modern cryptography, hinges on the difficulty of factoring large numbers. Specifically, it relies on the seemingly simple, yet computationally intensive, task of finding the prime factors of a composite number. The algorithm utilizes two large prime numbers, p and q, to generate a public key and a private key. The public key, which is made publicly available, is used to encrypt messages, while the private key, kept secret, is used to decrypt them. The security of RSA lies in the fact that even with the public key (which includes the product n = pq), it's computationally infeasible to factor n into its prime components p and q* within a reasonable timeframe, especially when dealing with extremely large numbers. This asymmetry between the ease of encryption and the difficulty of decryption is what makes RSA so effective. Understanding the properties of prime numbers and their unique distribution is thus crucial to grasping the underpinnings of RSA's security.

The Diffie-Hellman key exchange, a different approach to secure communication, offers a method for two parties to establish a shared secret key over an insecure channel. This protocol elegantly leverages modular exponentiation, which is deeply rooted in concepts like modular arithmetic explored earlier in this text. It begins by selecting a large prime number, p, and a generator g modulo p. Each party then chooses a secret integer, a and b, respectively, calculating their public keys using these values. These public keys are exchanged. The shared secret key is then derived using these exchanged public keys and the original secret values—each party independently computes the same value via modular exponentiation. Eavesdroppers only see the public values and the prime p and generator g; they can't compute the shared secret. The mathematical complexity of the discrete logarithm problem prevents the extraction of the secret keys from the public information, making the exchange secure. This method enables secure communication without ever explicitly transmitting the secret key itself. A fascinating aspect of cryptography is its reliance on unsolved problems in number theory. The security of RSA and Diffie-Hellman, for instance, is directly linked to the computational hardness of factoring large integers and solving the discrete logarithm problem. These problems, despite having been studied for centuries, remain computationally intractable for suitably large numbers, making them ideal foundations for robust cryptographic systems. This intertwining of theoretical mathematics and practical applications underscores the profound impact that number theory has on our modern digital world. Further research into these problems may lead to either stronger cryptographic techniques or the potential weakening of current ones, highlighting the constant evolution of cryptography in response to advancements in computational power and mathematical understanding. The pursuit of computationally hard problems, therefore, is not merely an abstract mathematical endeavor; it is a critical component of safeguarding digital security.

Cryptography is not just about protecting sensitive data; it's integral to securing the infrastructure of the internet and our digital lives. The HTTPS protocol, for example, which secures web browsing, relies heavily on RSA and other similar cryptographic techniques to ensure that communications between web browsers and servers remain confidential. Digital signatures, which verify the authenticity and integrity of digital documents, also utilize cryptographic algorithms based on number theory principles. The secure transmission of financial transactions, online communication, and virtually any other online activity, requires strong cryptographic systems, demonstrating the widespread and critical application of number theory in protecting our information. Understanding the mathematics underlying these systems is not just an academic pursuit but a key to understanding and participating in a secure digital environment.

In essence, the applications of number theory in cryptography are vast and pervasive, influencing various aspects of modern life. The mathematical foundations underpinning cryptographic algorithms are not only fascinating from a theoretical perspective but also crucial for ensuring the security of our increasingly interconnected world. While this chapter has provided only a glimpse into the rich tapestry of this connection, understanding the underlying principles of RSA, Diffie-Hellman and their mathematical basis is essential to appreciate the complexities and importance of cryptography in a digital age. Further exploration into the intricacies of number theory will enhance comprehension of the mechanisms that safeguard our digital information and communications.

6.1 RSA Encryption

RSA encryption is a cornerstone of modern cryptography, securing online transactions and communications daily. Its strength lies in the difficulty of factoring large numbers—a problem that has baffled mathematicians for centuries. At its heart, RSA relies on modular arithmetic and the properties of prime numbers, concepts we've already explored extensively.

The RSA algorithm uses two large prime numbers, p and q, to generate a public key and a private key. The product of these primes, n = p * q, forms part of the public key. This public key, along with another value calculated from p and q, is then made publicly available. Anyone can use this public key to encrypt a message.

The private key, however, is kept secret. It's derived from p and q through a specific calculation involving Euler's totient function, (n), which counts the positive integers less than or equal to n that are relatively prime to n. Importantly, calculating (n) is straightforward when you know p and q; however, determining (n) from just n is computationally infeasible for very large values of n. This computational asymmetry forms the foundation of RSA's security.

To encrypt a message, represented as a number m, we raise it to the power of the public exponent e, modulo n. The result, c me (mod n), is the ciphertext. Only someone possessing the private key can decrypt this message. Decryption involves raising the ciphertext to the power of the private exponent d, modulo n, yielding the original message: m cd (mod n). The values d and e are chosen such that e * d 1 (mod (n)). This congruence ensures that the decryption process correctly recovers the original message.

The security of RSA hinges on the difficulty of factoring n into its prime factors, p and q. If an attacker could efficiently factor n, they could calculate (n) and subsequently determine the private exponent d, compromising the encryption. Current algorithms for factoring large numbers are computationally expensive, rendering this approach impractical for sufficiently large values of n.

Choosing appropriate prime numbers is crucial for RSA's security. These primes must be sufficiently large to withstand modern factoring algorithms; otherwise, the encryption can be easily broken. Furthermore, they should be randomly selected to minimize the risk of finding predictable patterns that could aid in factorization. The security of RSA directly correlates with the size of n, typically requiring hundreds or even thousands of bits to guarantee sufficient protection.

Implementing RSA involves careful selection of parameters to achieve a balance between security and efficiency. Larger values of n provide higher security but also require more computational resources for encryption and decryption. Therefore, choosing the appropriate key size depends on the sensitivity of the data being protected and the computational capabilities available.

In summary, RSA encryption leverages the mathematical properties of prime numbers and modular arithmetic to provide a robust cryptographic system. Its effectiveness rests on the computational infeasibility of factoring large numbers. While the algorithm itself is relatively straightforward, its successful application requires careful attention to detail, particularly in selecting appropriate parameters to ensure the security of the encrypted data. Understanding the underlying mathematical principles is fundamental to appreciating the power and limitations of this widely used encryption method. The RSA algorithm, while powerful, should be combined with other security protocols and practices for comprehensive security in a real-world environment. Modern implementations often incorporate additional measures to address vulnerabilities and to improve overall system robustness. This includes regular key updates and the use of strong random number generators. Proper implementation is crucial to ensure its effectiveness. The significance of RSA lies not only in its mathematical elegance but also in its widespread adoption as a crucial component of secure communication. Its continued use highlights the practical impact of number theory on daily life.

6.2 Diffie-Hellman Key Exchange

The Diffie-Hellman key exchange is a revolutionary method for two parties to establish a shared secret key over an insecure channel, a cornerstone of modern cryptography. Its brilliance lies in its reliance on modular arithmetic and the discrete logarithm problem, a computationally hard problem forming the algorithm's security foundation. Imagine two individuals, Alice and Bob, wanting to communicate securely. They don't share any prior secret, yet need to agree on a key for encrypting their subsequent conversations. This is precisely where Diffie-Hellman steps in.

The process begins with the public selection of two large prime numbers, p, and a generator, g, a number less than p that generates all the numbers between 1 and p-1 using modular exponentiation. Both p and g are publicly known, forming part of the system's parameters. This public knowledge doesn't compromise security; the strength of the algorithm rests on the computational difficulty of solving the discrete logarithm problem, namely finding x given gx mod p. This is a computationally expensive process for sufficiently large primes.

Alice secretly chooses a random integer, a, and computes A = ga mod p. Bob, independently, chooses a random integer, b, and calculates B = gb mod p. Both A and B are then publicly exchanged, transmitted over an insecure channel; eavesdroppers can only access these values. The crucial point is that knowing A (or B) doesn't reveal a (or b) due to the one-way nature of modular exponentiation.

Now comes the crucial step of key establishment. Alice receives B from Bob and computes s = Ba mod p. Simultaneously, Bob receives A from Alice and computes s = Ab mod p. Remarkably, both Alice and Bob arrive at the same value, s, which is their shared secret key. The magic lies in the commutative property of modular exponentiation: s = Ba mod p = (gb)a mod p = gba mod p = (ga)b mod p = Ab mod p. They've successfully generated a shared secret without ever directly transmitting it.

The security of Diffie-Hellman relies on the difficulty of computing the discrete logarithm. Even though an eavesdropper has access to g, p, A, and B, calculating a or b—and thus s—is computationally infeasible with current technology for appropriately large p. Choosing p and g carefully is vital; weak choices can significantly reduce security. Using a known prime number may compromise the security of the encryption. The prime number needs to be a very large number that no computer will ever find a solution.

While exceptionally robust, Diffie-Hellman isn't without vulnerabilities. Man-in-the-middle attacks pose a significant threat. A malicious actor could intercept the exchange of A and B, substituting their own values, thus establishing separate shared secrets with both Alice and Bob, leaving them vulnerable to interception. This highlights the need for authentication mechanisms in conjunction with the Diffie-Hellman exchange to ensure the identities of communicating parties. Digital signatures are a common solution to mitigate this vulnerability. Such measures verify the authenticity of exchanged information, preventing man-in-the-middle attacks and significantly improving security. The combination of Diffie-Hellman with robust authentication safeguards is essential for securing modern communications.

In conclusion, the elegance and effectiveness of the Diffie-Hellman key exchange are undeniable, providing a robust foundation for secure communication in the digital age. Its mathematical underpinnings make it a fundamental component of modern cryptography, securing countless online transactions and communications daily. Understanding its inner workings, however, is crucial for appreciating its limitations and for employing it safely. The combination of careful parameter selection and supplemental security measures significantly boosts its effectiveness, enabling secure communication in an increasingly insecure environment. The Diffie-Hellman key exchange is, therefore, more than just an algorithm; it's a testament to the power of number theory in protecting our digital world.

7 Further Explorations

Let's delve into the complexities beyond linear Diophantine equations. We've covered the basics; now, consider the challenge of solving equations like $x^3 + y^3 = z^3$. This seemingly simple extension leads to surprisingly intricate mathematical landscapes. The absence of non-trivial integer solutions, famously proven by Andrew Wiles as a culmination of decades of mathematical effort, is a testament to the profound depth hidden within seemingly straightforward problems. Understanding this requires a deep dive into advanced algebraic techniques, far beyond the scope of a basic introduction, but it highlights the rich interconnectedness of mathematical concepts. The journey from simple linear equations to this cubic equation exemplifies the escalating complexity one encounters as you progress in number theory. The proof itself is a marvel of modern mathematics, weaving together various areas like elliptic curves and modular forms to reach its conclusion. This underscores the rewarding nature of persistent exploration in this fascinating field.

Moving beyond Diophantine equations, we can explore elliptic curves. These are curves defined by equations of the form $y^2 = x^3 + ax + b$, where a and b are constants. They might appear simple at first glance, but they possess a rich structure and profound connections to other areas of mathematics. Their study involves considering points on the curve and how they can be added to each other, creating a group structure. This seemingly abstract operation has far-reaching consequences. Elliptic curves form the basis of modern cryptographic systems, offering a level of security unmatched by simpler methods. The properties of these curves are crucial for understanding these cryptographic applications, ensuring the security of online transactions and sensitive data. This demonstrates the power of seemingly theoretical mathematics in securing our digital world.

Finally, let's consider number fields. Until now, we've largely focused on rational numbers – fractions of integers. However, number theory extends far beyond this familiar realm. Number fields involve expanding our number system to include solutions to polynomial equations with integer coefficients. For example, the equation x^2 \item 2 = 0 has no rational solution, but it does have solutions in the real numbers (2 and -2). The smallest field containing these solutions and the rational numbers is a number field, extending our mathematical landscape significantly. The concept opens doors to a more general and abstract study of number theory, expanding the scope of investigations. The study of algebraic numbers – roots of polynomials with integer coefficients – within these number fields brings an entire new level of complexity and abstraction to the field, allowing us to extend many concepts and theorems we've already learned to this wider setting. Such exploration further demonstrates the power and beauty of abstract mathematical frameworks. Exploring these areas provides a strong foundation for advanced studies in number theory. You'll encounter new challenges and rewards as you further your mathematical journey, unlocking deeper insights into the fundamental structures of numbers. The elegance and power of these concepts are apparent only after dedicated effort, which is greatly rewarded upon mastering the material.

7.1 Advanced Diophantine Equations

Moving beyond linear Diophantine equations and Pythagorean triples, we encounter a vast landscape of more intricate problems. Consider Fermat's Last Theorem, a seemingly simple statement with a profoundly complex proof: no three positive integers a, b, and c can satisfy the equation a + b = c for any integer value of n greater than 2. Its solution, achieved by Andrew Wiles after centuries of efforts, involved highly advanced mathematical tools far beyond the scope of this introductory text, but understanding its fundamental statement lays a crucial foundation for grasping the challenges inherent in higher-order Diophantine equations.

The equation $x^2 + y^2 = z^2$, the basis for Pythagorean triples, is a specific case of a much broader class of equations known as elliptic curves. These curves, represented by the general form $y^2 = x^3 + ax + b$, hold immense significance in modern number theory and cryptography. Their study involves concepts like group structure, modular forms, and the deep connections between seemingly disparate branches of mathematics. Solutions to elliptic curves often exhibit intricate patterns, and understanding these patterns has profound implications. For example, the famous Birch and Swinnerton-Dyer conjecture, still unproven, proposes a link between the number of rational points on an elliptic curve and the behavior of an associated L-function.

Beyond elliptic curves, the realm of Diophantine equations expands into higher dimensions and involves polynomial equations with multiple variables. These equations often present formidable challenges, with solutions often requiring sophisticated techniques from algebraic geometry and number theory. The difficulty lies not just in finding solutions, but also in determining whether solutions exist and how many there are. Determining the existence of solutions might involve proving the solvability of a system of polynomial congruences modulo various primes, using tools such as Hensel's lemma for lifting solutions from modulo p to modulo p. This method is valuable in solving higher-degree Diophantine equations, systematically building a solution from local solutions.

A key area of study involves exponential Diophantine equations. These involve equations where the variables appear as exponents, often leading to intricate patterns and connections to areas like transcendental number theory. For instance, the Catalan conjecture, now proven as Mihailescu's theorem, states that the only solution in integers to the equation $x \quad y = 1$, with x, y, m, and n being integers greater than 1, is $3^2 \quad 2^3 = 1$. The proof of this, which wasn't simple, relied on sophisticated techniques like cyclotomic fields and linear forms in logarithms.

Another significant class includes Thue equations, which take the form

$f(x,y) = k$, where $f(x, y)$ is an irreducible homogeneous polynomial of degree at least 3. Solutions are surprisingly sparse, even for relatively low degrees, often involving deep results from algebraic number theory and the geometry of numbers. Understanding the bounds on the size of solutions to these equations represents a significant undertaking. Determining bounds and properties of solutions to these types of equations is an area of active research. Thue's theorem, a fundamental result in this area, guarantees that Thue equations have only a finite number of integer solutions, a cornerstone in the study of Diophantine equations.

Finally, it is worth noting that many Diophantine equations remain unsolved, posing significant challenges for mathematicians. The exploration of these advanced equations represents a frontier of mathematical research, continuously pushing the boundaries of our understanding of numbers and their relationships. The ongoing study of these problems provides constant insights into the fundamental structure of number systems and their inherent properties, making advanced Diophantine equations a vital area in number theory. The quest for solutions often leads to the development of new mathematical techniques and theorems, enriching the field and driving further exploration. The study of these seemingly abstract problems can surprisingly lead to unexpected applications in various fields, highlighting the interconnectedness of mathematics.

7.2 Elliptic Curves

Elliptic curves, deceptively simple in their equation form, unlock profound mathematical depths. Defined by the equation $y^2 = x^3 + ax + b$, where a and b are constants and the discriminant $4a^3 + 27b^2$ is nonzero to avoid singularities, these curves provide a fertile ground for exploring intricate properties. The non-singularity condition ensures the curve is smooth, without cusps or self-intersections, a crucial property for many of its applications. This seemingly straightforward definition belies the richness of their structure and their surprisingly widespread applications, extending far beyond the realm of pure mathematics.

Consider the points on the curve; these are the solutions (x, y) that satisfy the equation. The fascinating aspect of elliptic curves lies in their ability to form an abelian group. Imagine adding two points on the curve. This isn't simple addition of coordinates. Instead, a geometric construction is involved: draw a line through the two points; it will intersect the curve at a third point. Reflecting this third point across the x-axis gives the sum of the original two points. The addition operation satisfies group axioms, demonstrating a profound connection between geometry and algebra. This group structure provides a powerful tool for investigating these curves' mathematical properties. Understanding this group law is fundamental to applying elliptic curves in cryptography and other areas.

For a beginner, grasping this geometric addition might seem challenging, but visualizing it helps immensely. Start by plotting some points on a graph representing the curve. Pick two points and observe the line that connects them. If you extend that line, it'll intersect the curve again. Then, you visualize the reflection about the x-axis and you are done. If we have a point and itself, the process is a bit different: instead of connecting the two points, we consider the tangent line to the curve at the single point we are given. This tangent line intersects the curve again, and the reflection of this intersection point about the x-axis is our 'doubled' point. This geometric interpretation makes the algebraic representation more intuitive. Practice with different examples.

The beauty of elliptic curves extends to their application in cryptography. The discrete logarithm problem, crucial in securing many cryptographic systems, becomes significantly harder to solve on elliptic curves, resulting in better security with smaller key sizes compared to traditional multiplicative groups. The difficulty of the discrete logarithm problem stems from the complex nature of the group operation on elliptic curves. It is significantly harder to calculate a multiple of a point (which can be achieved simply by repeating the geometric addition) than determining which point, when multiplied by an integer, produces a specific result. This inherent complexity makes elliptic curve cryptography (ECC) a cornerstone of modern secure communication.

Furthermore, elliptic curves are deeply intertwined with number theory. Points on an elliptic curve with rational coordinates form a group. Exploring these rational points involves intricate Diophantine equations and opens doors to profound results. For instance, investigating the number of rational points on specific elliptic curves can lead to complex and challenging problems, some of which remain unsolved. This connection to Diophantine equations and integer solutions further underscores the relevance of elliptic curves within the broader framework of number theory. This shows that even seemingly simple equations contain profound mathematical depth.

In conclusion, elliptic curves offer a fascinating blend of geometry and algebra, with significant applications in various fields. Their group structure, the difficulty of the discrete logarithm problem on them, and their connection to Diophantine equations highlight their significance in both pure and applied mathematics. Understanding the basic concepts of elliptic curves, including the geometric addition of points and the significance of their group structure is a crucial step in appreciating the vast implications of this area of mathematics. While the deeper intricacies require further study, grasping the fundamentals provides a strong foundation for exploring this remarkable area further. Their practical applications in cryptography ensure their continued relevance in securing our digital world. Therefore, continued exploration of elliptic curves promises rich rewards in both theoretical and practical advancements. Further exploration of this topic may involve more detailed discussion of algorithms, particularly for point multiplication on elliptic curves, a critical computation within ECC.

7.3 Number Fields

Number fields represent a significant expansion of the familiar number systems we've explored. Instead of just dealing with rational numbers, which are ratios of integers, we now consider fields containing rational numbers and also other algebraic numbers. These algebraic numbers are roots of polynomial equations with rational coefficients. Think of it as building upon the foundations we've already established, taking our understanding to a higher level of abstraction and complexity, revealing a richer mathematical landscape. The elegance and power lie in their ability to solve previously intractable problems and to provide new avenues for mathematical exploration.

Consider a simple quadratic equation like x^2 \item 2 = 0. Within the rational numbers, this equation has no solution. However, if we extend our number system to include the square root of 2, we suddenly have solutions. This extended system, containing all numbers of the form a + b2 where 'a' and 'b' are rational, forms a number field, often denoted as Q(2). This is just the beginning, the tip of the iceberg. We can create many such fields, each possessing unique properties and characteristics that can be analyzed with the tools and concepts we've developed earlier in the book. Understanding these properties opens doors to solving problems previously considered unsolvable.

Constructing these number fields involves examining polynomials. Every number field can be generated by adjoining a root of an irreducible polynomial with rational coefficients to the field of rational numbers. This process of adjoining a root essentially "expands" our number system, incorporating new elements that satisfy the polynomial equation. The degree of the polynomial, which dictates the highest power of x in the equation, determines important properties of the corresponding number field. A polynomial's degree directly impacts the field's structure and the complexity of the arithmetic within it. These relationships are fundamental to understanding the behavior and properties of number fields.

The concept of an "ideal" is particularly crucial when working with number fields. Ideals are subsets of the number field ring that behave well under addition and multiplication. Unlike previous concepts, ideals aren't necessarily sets of numbers; they are sets with particular multiplicative properties. This abstract notion becomes essential for analyzing the structure of the field, helping to unravel its intricate inner workings and understand its relationships to other fields. This is a departure from the more concrete numbers we've dealt with thus far, necessitating a shift in perspective and a grasp of abstract algebraic concepts.

Furthermore, the arithmetic within a number field can be significantly different from the familiar arithmetic of rational numbers. For instance, unique factorization, a property we've relied upon extensively, might not hold true in all number fields. This means a number might have different factorizations into "irreducible elements" – the field's equivalent of prime numbers – leading to significant challenges and a need for innovative approaches to factorization problems. This difference emphasizes the need for flexible thinking and the adaptation of established techniques. The properties of unique factorization are not guaranteed in the expanded realm of number fields, requiring new techniques and approaches to handle factorization.

One way to address the lack of unique factorization is to introduce the concept of ideal factorization. It allows us to restore a sense of unique factorization, albeit on a different level. Instead of factorizing numbers directly, we factor ideals into prime ideals, obtaining a unique factorization. This is a powerful tool, allowing us to analyze the structure of number fields even in cases where the usual prime factorization fails. This shift from elements to ideals enables us to regain a sense of order and predictability in the number field's arithmetic structure.

Number fields have profound implications for various areas of mathematics, including algebraic number theory, and even connect to the study of Diophantine equations that we previously explored. By embedding Diophantine equations within the context of number fields, we can analyze them from a different perspective, often leading to more efficient methods of solution or a deeper understanding of their inherent properties. This interdisciplinary connection emphasizes the richness and interconnectedness of different areas of mathematical study. Their applications extend far beyond mere theoretical curiosity, offering powerful tools for analyzing complex mathematical structures and addressing challenging problems across numerous disciplines. The study of number fields demonstrates a crucial bridge between seemingly disparate areas of mathematical study, showcasing the unifying power of abstract mathematical concepts.

The study of number fields, although abstract, offers a deeper understanding of the fundamental nature of numbers. It builds upon the concepts we have previously learned, allowing for a more profound and complete appreciation of the rich structure and complexity inherent within the system of numbers. By moving beyond the rational numbers, we encounter fascinating phenomena and powerful tools that enhance our ability to solve problems and explore previously uncharted mathematical territories. Number fields aren't simply an extension; they represent a significant leap forward in the understanding of number theory's breadth and depth.

www.ingramcontent.com/pod-product-compliance
Lightning Source LLC
LaVergne TN
LVHW022016200926
848481LV00007B/2421